PREFACE

The U.S. Department of Energy (DOE) is charged under the Energy Independence and Security Act of 2007 (EISA 2007) with modernizing the nations electricity grid to improve its reliability and efficiency. As part of this effort, DOE is also responsible for increasing awareness of our nations Smart Grid. Building upon *The Smart Grid: An Introduction*, a DOE sponsored publication released in 2008 and available online at www.smartgrid.gov, this publication is one in a series of books designed to better acquaint discrete stakeholder groups with the promise and possibilities of the Smart Grid. Stakeholder groups include Utilities, Regulators, Policymakers, Technology Providers, Consumer Advocates and Environmental Groups.

TABLE OF CONTENTS

SECTION 01 // PAGE 2
The Case for Modernization: *Why the future won't can't be like the past.*

SECTION 02 // PAGE 5
The Smart Grid: *Benefits to utilities.*

SECTION 03 // PAGE 10
Rates & Regulations: *Possible approaches.*

SECTION 04 // PAGE 13
Standards & Security: *Getting to certainty.*

SECTION 05 // PAGE 14
Smart Grid & the Environment: *Enabling a cleaner energy future.*

SECTION 06 // PAGE 18
The Smart Grid Progress Report: *Who's doing what out there?*

SECTION 07 // PAGE 20
The Smart Grid Maturity Model: *Because one size doesn't fit all.*

SECTION 08 // PAGE 22
FERC, NARUC & the Smart Grid Clearinghouse: *Drawing clarity from complexity.*

SECTION 09 // PAGE 24
Next Steps: *Summoning the energy.*

GLOSSARY // PAGE 26
Smart Grid terms worth knowing.

RESOURCES // PAGE 27
Places to go to learn more.

> *Nationwide, demand for electricity is expected to grow 30% by 2030. Electricity prices are forecast to increase 50% over the next 7 years.*

THE CASE FOR MODERNIZATION: WHY THE FUTURE ~~WON'T~~ CAN'T BE LIKE THE PAST.

Our electrical grid is a certified wonder of the world. For more than a century, the United States electrical power grid has been reliable, efficient and worthy of its designation by the National Academy of Engineering as the engineering marvel of the 20th century.

UTILITY: A DEFINITION

According to the Energy Information Administration, an electric utility is "any entity that generates, transmits, or distributes electricity and recovers the cost of its generation, transmission or distribution assets and operations, either directly or indirectly, through cost-based rates set by a separate regulatory authority (e.g., State Public Service Commission), or is owned by a governmental unit or the consumers that the entity serves. Examples of these entities include: investor-owned entities, public power districts, public utility districts, municipalities, rural electric cooperatives, and State and Federal agencies."

Historically, one could argue that its inherent business model has been powered in part by its predictability. To paraphrase one utility executive, for 100 years customers have been trained not to worry about the price of electricity. For roughly the same amount of time, utility executives, understandably trained for risk aversion, have been governed by the irresistible urge to leave well enough alone.

As of now, this landscape has irrevocably changed. Facing a future in which unpredictability will be the norm, the grid's signature strength is suddenly its weakness. Clearly, these are times that call for a smarter grid, however you choose to define it.

Consider just a few of the drivers:

The costs of new generation and delivery infrastructure are climbing sharply. According to The Brattle Group — a consulting group that specializes in economics, finance, and regulation — investments totaling approximately $1.5 trillion will be required over the next 20 years to pay for the infrastructure alone.

Nationwide, demand for electricity is expected to grow 30% by 2030.[1]

Electricity prices are forecast to increase 50% over the next 7 years.[2]

Spiraling electricity rates and the cost of carbon (to be fully ascertained through the outcome of proposed cap-and-trade legislation) are combining to reveal the true — i.e., higher — cost of energy.

Meanwhile, a changing consumer is already reshaping interaction at the meter. Research is incomplete as to how much control over

their energy choices customers ultimately will seek to exercise. Yet their awareness has been heightened by projects large and small, from the proliferation of Advanced Metering Infrastructure (AMI) projects to high-profile developments in states such as Texas, California, Colorado and Hawaii. And if their recent telecommunications history is any guide or the so-called Prius effect to be believed, customers will be demanding more control rather than less. Just tell them what they're paying for (or how they might be able to pay less) and watch what happens.

Apart from cost considerations, recent polls indicate that 75% of Americans support federal controls on the release of greenhouse gases in an effort to reduce global warming, 54% "strongly." Even among those who are "very" concerned about the cost impact, two-thirds support the regulation.[3]

Whether or not we call it the Smart Grid, the industry is organically moving toward modernization, with more distributed generation in the form of smaller generators, more customer interaction, the integration of more variable resources such as wind and solar, and more renewables overall. (The top four renewable technologies display a growth rate of more than 20% per year. Worldwide annual investment topped $70 billion in 2007.[4])

WHY YOU NEED TO BE A SMART GRID FACILITATOR

For utilities, the job is the same as it ever was — the exercise of responsive and responsible control. With thoughtful adoption of the Smart Grid and the overlay of new tools, techniques and technologies, the "big picture" gets clearer. Utilities will be able to view and measure what's going on in the system more completely and more frequently than is currently possible, which in turn will enable additional levels of control.

Such control will allow utilities to better optimize the grid to support a number of public policies, from intermittent renewable integration to more dynamic interfaces with customers. This will also offer utilities more flexibility relative to how they use energy toward the greater societal objectives of reducing greenhouse gases and energy consumption.

At a minimum and in the shorter term, a smarter grid offers utilities operational benefits (outage management, improved processes, workforce efficiency, reduced losses, etc.) as well as benefits associated with improved asset management (system planning, better capital asset utilization, etc.).

THE PRIUS EFFECT

The Prius makes a strong anecdotal case for "letting the customer drive" when it comes to energy decisions. Toyota's most renowned hybrid vehicle features a dashboard monitor that constantly indicates what effect your driving habits have on your efficiency and makes visible — in real-time — the consequences of your energy usage. The resulting "Prius Effect" has been cited by various energy and computing researchers as convincing evidence that consumers will readily change their habits if exposed to feedback in real time.

3

THE CHALLENGE FROM HERE:
HOW TO ADDRESS THE SMART GRID

It is an article of faith that grid infrastructure is beginning to fail us. It is a given that modernizing this infrastructure will be extremely costly for all of us. The question is: Do we make these investments simply by relying on what has gone before or commit to the forward-looking organizing principle known as the Smart Grid?

Consider this a prospectus on the potential of our present and future grid. You'll learn about the barriers and opportunities relative to Smart Grid adoption; you'll discover how some utilities have already taken significant steps or put projects in place; you'll see how consensus is being achieved as various stakeholders align behind the need for a Smart Grid, if not exactly agreeing on the steps needed to get there. Finally, you'll learn about a tool you can use to assess your utility's Smart Grid appetite and readiness.

Within the following pages, you will see why you should incorporate the Smart Grid into your utility's business plan. As a Smart Grid "accelerator," you can play a critical role in modernizing the grid and benefiting your utility.

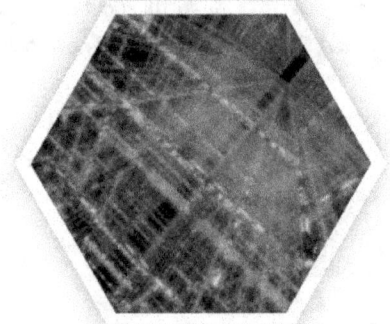

WHAT IS THE SMART GRID?

● *Defining the Smart Grid is in itself tricky business. Select six stakeholders and you will likely get at least six definitions.*

For utilities, it is this:

The Smart Grid is the electric delivery network from electrical generation to end-use customer, integrated with the latest advances in digital and information technology to improve electric-system reliability, security and efficiency.

SMARTER GRID / SMART GRID

● *Because it is deploying now yet will only be fully realized over time, it is necessary to split one Smart Grid into two for the purpose of discussion: A smarter grid refers to the current state of the transformation, one in which technologies are being deployed today or in the near future. The Smart Grid is the ultimate vision — the full realization of everything it can be.*

With real time data made possible by Smart Grid technologies, utilities will be able to more effectively utilize assets under normal and adverse conditions.

THE SMART GRID: BENEFITS TO UTILITIES.

When viewed relative to "the grid we have now," transformation to a smarter grid will lead to enhancements that will positively affect every aspect of electricity generation, delivery and consumption, as most recently detailed by DOE's Modern Grid Strategy and the Electricity Advisory Committee.

THE HIGHLIGHTS...

● The Smart Grid will increase the overall use and value of existing production and transmission capacity; incorporate greater levels of renewable energy; reduce carbon emissions by increasing the efficiency of the system; gain functionality out of increasing energy intensity; improve power quality to correspond to new digital demands; and do it all with the highest levels of security.

What's more, the ability of a utility to build a smarter grid incrementally rather than all at once is looked upon far more favorably by financial markets.

OPTIMIZING ASSET UTILIZATION AND EFFICIENT OPERATION

In 2005, excluding fuel and purchased power, investor-owned utilities spent $40 billion to operate and maintain the power system. With real-time data made possible by Smart Grid technologies, utilities will be able to more effectively utilize assets under normal and adverse conditions. Among the benefits: A reduction in failure-related maintenance and outage costs and a longer service life among some of the assets. Overall and over time, integrated communications technologies will lessen the need for new and costly hard assets.

ENHANCING RELIABILITY

The Smart Grid will dramatically reduce the cost of power disturbances. Communications and control technologies applied to the grid will be able to isolate faults and allow more

rapid restoration of service. Decision-support systems will "know" when there is the need to quickly reduce load and respond autonomously to adverse conditions.

The Smart Grid will also be able to "call for help," enlisting support from distributed energy resources to help balance system needs.

In combination, such functionality will strengthen the transmission and distribution system, increase operational flexibility and greatly reduce the risk of a failure that might affect the entire grid.

POINT OF CLARIFICATION: WHAT THE SMART GRID ISN'T

It's only natural to confuse the terms Smart Grid and smart meters. The general news media do it all the time. But smart metering and the physical meter itself are just examples of a single enabling technology that makes two way communication possible.

IMPROVING POWER QUALITY

Power quality events — dips in voltage lasting less than 100 milliseconds — can have the same effect on an industrial process as a more general outage that lasts minutes. A single such event can cost commercial facilities such as banks and data centers millions of dollars.

According to the Electric Power Research Institute (EPRI), by 2011, fully 16% of our nation's electric load will require digital-quality power. (And digital equipment is far more sensitive than analog ever was, requiring tighter tolerances for voltage and frequency fluctuation.) The Smart Grid will help limit the impact of power-quality events. Transmission-side Smart Grid components will work to reduce voltage sags and swells. On the distribution level, disturbed sources could be removed and replaced with clean backup power supplies.

Broad-based power-quality improvements will reduce losses to American businesses across the board, from scrapped materials in industrial processes to the number of lost customers in a retail environment.

REDUCING WIDESPREAD OUTAGES

A $10-billion event

According to the "Final Report on the August 14, 2003 Blackout in the United States and Canada," that was the estimated price tag for our nation's last massive blackout, which left

more than 28 million people in Michigan, New York and Ohio living without power for up to 4 days. Already, "lessons learned" from this event have resulted in a smarter grid and the institution of enforceable reliability standards.

That said, the Smart Grid will be able to employ multiple technologies to ensure that such a scenario is not repeated. Improved interfaces and decision-support tools will enable system operators to monitor the status of the grid at a glance — detecting threats against it — and identify, relieve and/or replace failing equipment even before a breakdown can occur. In some cases, power-stabilization software will be able to address an event and "heal" faster than humans can react. Even grid-friendly appliances will play a role, responding to demand-response signals to adjust load.

REDUCING VULNERABILITY TO MADE-MADE EVENTS AND NATURAL DISASTERS

Overlaying the entire electrical network, the Smart Grid's integrated communications infrastructure will provide detection and mitigation of both cyber and physical threats. Its ability to support a wide variety of generation options also reduces the effects of an attack at any one point on the system. Indeed, its strength is in its diversity. For example, whether natural or man-made, a diversity of distributed energy resources offers grid operators a variety of options in response to an emergency.

10/28
10 BILLION
Dollars
28 MILLION
People

Similarly, resource diversity within a geographic region offers additional means to restore the grid, and a diversity of fuels increases the likelihood that adequate power will be available.

IMPROVING PUBLIC AND WORKER SAFETY

According to the American Public Power Association, utility work is among the most dangerous occupations, resulting in 1000 fatalities and 7000 flash burns annually. Rapid identification of problems and hazards made possible by improved monitoring and decision-support systems will be able to predict equipment failure before it occurs to save lives and reduce injuries. Clearly, it is easier to service equipment routinely than during an outage event. Reducing failures also leads to reducing outages, which means traffic lights, elevators, etc., continue to function for the benefit of the public's safety.

IMPROVED ECONOMICS

Efficiencies ushered in by the Smart Grid should mitigate some of the rising costs of electricity. Real-time price signals will allow consumers to participate in the electricity market based on current supply and demand pricing scenarios. Communication among these buyers and sellers should reduce grid congestion and unplanned outages, as well as determine the real price for electricity at various times throughout the day. The reach of market efficiencies is also improved. The analyst group LECG recently determined that the organized wholesale electricity markets of PJM and the New York Independent System Operator (ISO) have already reduced average wholesale electric rates between $430 million and $1.3 billion a year.

MORE ROBUST MARKETS

The Smart Grid will help welcome new market participants, enabling a variety of new load management, distributed generation, energy storage and demand-response options and opportunities. These contributions are

Smart Grid **Key Technology Areas**

Integrated Two-Way Communications make the Smart Grid a dynamic, interactive, real-time infrastructure. An open architecture creates a plug-and-play environment that securely networks grid components and operators, enabling them to talk, listen and interact.

Advanced Components play an active role in determining the electrical behavior of the grid, applying the latest research in materials, superconductivity, energy storage, power electronics, and microelectronics to produce higher power densities, greater reliability and power quality.

Advanced Control Methods monitor power system components, enabling rapid diagnosis and timely, appropriate responses to any event. Additionally, they also support market pricing, enhance asset management and efficient operations.

Sensing and Measurement Technologies enhance power system measurements and facilitate the transformation of data into information to evaluate the health of equipment, support advanced protective relaying, enable consumer choice and help relieve congestion.

Improved Interfaces and Decision Support will enable grid operators and managers to make more accurate and timely decisions at all levels of the grid, including the consumer level, while enabling more advanced operator training.

BENEFITS FOR COMMERCIAL AND INDUSTRIAL CUSTOMERS

● Electric motors consume approximately 65% of industrial electricity, understandable because they power virtually every process necessary for moving things from compressed air to conveyor belts. Variable-speed drives can reduce a motor's energy consumption by up to 60% compared with fixed drives and can be enabled to respond to a utility's price signals. Imagine the impact that such communication can have on manufacturing specifically and society in general.

reinforcing the Smart Grid's economic advantages by allowing demand to act as a supply resource, enabling utilities to defer some large capital investments in power plants, substations and transmission and distribution lines. As a result, tens of billions of dollars will be saved over a 20-year period, according to the Pacific Northwest National Laboratory. By increasing the grid's robustness and efficiency, options such as these will work to reduce peak prices and demand, leading to cost savings and downward pressure on rates for all stakeholders.

Demand response is already illuminating the promise of the Smart Grid through greater enablement in certain regions of the country. Demand response is a means by which demand will be dynamically and continuously balanced with supply-side resources to reduce price volatility. Distributed energy resources (DER) may accelerate consumer usage of small generation and storage devices through connections with grid and two-way flows of electricity and communications.

MORE ENVIRONMENTALLY FRIENDLY

In enabling the deployment of all forms of generation and storage, the Smart Grid will encourage greater use of distributed energy resources, including maximizing the use of existing combined heat and power (CHP) units. Residing primarily at large commercial and industrial sites, existing

CHP units — the CO_2 emissions profile of which are substantially lower than fossil-fueled power plants — represented 83.5 gigawatts (GW) of installed capacity as of 2005. DOE estimates suggest that additional opportunities could be as high as 130 GW.[5]

In being able to access a wider diversity of fuels, the Smart Grid will be able to generate more energy from carbon-free sources such as centralized hydro, wind, solar and nuclear power. In addition, it will be able to better take into account the intermittency of renewables.

Through the use of low-emission DER sources, the Smart Grid will enable states to more rapidly approach their Renewable Portfolio Standards (RPS) goals (See page 17).

REDUCTION IN ELECTRICAL LOSSES

Electrical generation is required to "cover" system losses; that is, for the system to work, power is required to provide the energy consumed by line loss and inefficient equipment. Smart Grid components and other efficiency improvements engineer this waste out of the system. With more generation alternatives at its disposal, the Smart Grid will be able to utilize many more near load centers and minimize transmission losses.

The Smart Grid increases opportunities for consumer choice while reducing the cost of delivered electricity. It makes firm the promise of clean, renewable energies such as wind and solar available at meaningful scale. It allows for the connection of an entire portfolio of resources. And it enables communication among all parties.

Yet it's important to remember that the Smart Grid is a journey rather than a destination. Through modernization efforts, a smarter grid will evolve into the fully integrated Smart Grid over time. And, much like every major modernization effort in history, it will face hurdles.

Consider the business case for investing in the Smart Grid. Utilities such as Austin Energy have proved the cost-effectiveness of its multi-dimensional Smart Grid investment. Currently, however, business cases for investing in the Smart Grid processes and technologies are often incomplete when viewed strictly with regard to near-term cost-effectiveness.

Invariably, it is easier to demonstrate the value of the end point than it is to make a sound business case for the intermediate steps to get there. Societal benefits, often necessary to make investments in modern grid principles compelling, are normally not included in utility business cases. Yet credit for those very societal benefits in terms of incentives and methods for reducing investment risks might stimulate the deployment of modern grid processes and technologies.

As study after study indicates, the societal case for Smart Grid adoption is fundamental, lasting and real:

> Increasing energy efficiency, renewable energy and distributed generation would save an estimated $36 billion annually by 2025.[6]

> Distributed generation can significantly reduce transmission-congestion costs, currently estimated at $4.8 billion annually.[7]

> Smart appliances costing $600 million can provide as much reserve capacity to the grid as power plants worth $6 billion.[8]

> Over 20 years, $46 billion to $117 billion could be saved in the avoided cost of construction of power plants, transmission lines and substations.[9]

WHAT THE SMART GRID CAN'T DO

🔘 *The Smart Grid cannot be reduced to a line item in a utility rate case, a fact that is both a weakness and a strength. Rather than presenting a compelling present-day justification for cost-allocation, it instead gives rise to societal benefits that in time will turn into system benefits. It therefore merits consideration in every utility's business plan.*

To fully capitalize upon grid modernization, elements of the Smart Grid plan must be as thoughtful as the technologies deployed.

RATES & REGULATIONS: POSSIBLE APPROACHES.

Currently, the benefits of the Smart Grid are not as apparent to many stakeholders as they could or should be. Like the early days of construction of the interstate highway system, it may be difficult to envision the Smart Grid's ultimate value during its building phase. In fact, perhaps all that certain observers can see when they consider the Smart Grid is disruption of the status quo.

DYNAMIC IDAHO

 *The Idaho Public Utilities Commission is actively gauging the effectiveness of dynamic pricing strategies. The state's time-variant pricing programs include Energy Watch, a simplified critical peak pricing program that rewards customers for reducing demand during summertime "Energy Watch events"; and a Time-of-Day program for customers who shift consumption of electricity from daytime hours to the late evening and weekends. Among the Commission's findings are that customers substantially reduced load during Energy Watch events.*

The state is also one of the "early adopters" of decoupling. A three-year pilot has been instituted and is currently deployed by the Idaho Power company. For a map of current decoupling activity by state, visit the website of the Institute of Electric Efficiency (IEE).

What is abundantly clear is that our costs are rising, our environment is suffering, our energy resources are finite — and we need a plan, disruptive or not. Try to imagine the interstate highway system without one: "Roads to Nowhere," everywhere. Or the Internet without an organizing principle. Millions might have access to e-mail, but millions more would be staring at blank screens.

To fully capitalize upon grid modernization, certain elements of the Smart Grid plan must be as thoughtful as the technologies deployed. Here, we enumerate a number of approaches towards that objective.

DYNAMIC PRICING

The typical electric bill of decades past was undecipherable to many and delivered long after the electricity was. Worse yet, that bill is still being snail-mailed today to far too many consumers of electricity, at a time when existing and emerging technologies make it possible for consumers to see the day-to-day cost of electricity. The capability of Advanced Metering Infrastructure (AMI) to facilitate two-way communication, interval metering and time-based billing make dynamic pricing an option for all classes of utility customers — including lower-income customers.

Dynamic pricing reflects hourly variations in retail power costs, furnishing customers the detail necessary to manage their utility bills in a variety of beneficial ways. Three principal categories of dynamic pricing include:

• *Real-time pricing* – rates are based on hourly fluctuations in wholesale markets, which enable consumers to plan their electric use to coincide with low prices.

• *Peak-time rebate* – the traditional blended rate applies, but customers can realize healthy rebates for reducing load during peak periods.

• Critical-peak pricing – prices can increase by 500% during peak periods, limited to a small number of hours per year. Customers agreeing to reduce usage in such hours will pay slightly lower rates for the remainder of the year.

Especially in the prevailing economy, customers may want to avail themselves of as many tools and choices as possible to control their usage and energy bills. Judiciously structured and applied, dynamic rates stand to benefit every consumer of electricity. Consider a working family, out of the house for most of the day with the kids at school. The family's ability to save money by participating in demand-response programs during the afternoon peak could be very beneficial to them.

Whether or not they choose to participate in a dynamic-rate program, dynamic rates reduce bills for all customers. There are savings to the system and ratepayers as a whole every time peak demand is reduced because the utility doesn't have to buy expensive power at 2 in the afternoon on July 15 or fire up that expensive peaking plant.

While some will argue that AMI is not cost-effective, recurring savings such as these can be put to work reinforcing the business case for AMI by offsetting the costs of AMI deployments and associated Smart Grid modernization.

INCENTIVIZING UTILITIES

The pros and cons of retail rate reform with respect to the Smart Grid include a number of hot topics. For example, historically a utility's rate of return has been based on the amount of power it generates and energy it sells. Absent in this model is the incentive for any party to conserve energy, which effectively leaves a utility's incentive to engage in demand response, energy efficiency and distributed generation out of the conversation. One way being proposed to redress this issue is decoupling.

Decoupling lowers a utility's rate of return because that utility is assuming less risk. In fact, since it now has certainty, it lowers the revenue requirements overall that customers otherwise would have to pay. If the utility over-recovers, it refunds the surplus to customers in the same way that if it under-recovered, it would require customers to pay a surcharge. Decoupling also brings a degree of transparency to rate cases among all parties – utilities, regulators and consumer advocates.

Some believe that such an incentive to save energy may make it more likely to subscribe to demand-response, energy-efficiency and distributed-generation programs that haven't "paid off" in the past.

EFFICIENCY ORGANIZATIONS: AN ALTERNATIVE APPROACH TO RETAIL RATE REFORM

NARUC holds the position that taking utilities out of the efficiency business and having that function played by a State, quasi-State, or private sector entity is a proven alternative to removing disincentives to their promoting efficiency. In fact, numerous examples exist of successful efficiency programs being delivered by non-utility providers. Examples of such organizations include Efficiency Vermont and the New York State Energy Research and Development Authority (NYSERDA).

Other stakeholders maintain that decoupling is not the answer, that it guarantees earnings to a utility rather than gives it the opportunity to earn. In response, decoupling advocates argue that it is precisely in removing this risk or uncertainty that enables utilities to take advantage of saving energy rather than generating even more of it.

NET METERING

Net metering programs serve as an important incentive for consumer investment in distributed energy generation, enabling customers to use generation on their premises to offset their consumption by allowing their electric meters to turn backwards when they generate electricity in excess of their demand. In some states, this offset means that customers receive retail bill credits for the electricity they generate themselves, rather than buy from the system.

THESE APPROACHES ARE NOT SELF-EVIDENT

It will require significant educational outreach to ensure that consumers and utilities alike understand the potential benefits that can be gained from decoupling, dynamic pricing, net metering and similar concepts as they apply to the Smart Grid. DOE is charged with raising their awareness. This book is just one of the many resources you have at your disposal; others are noted in the Resources section.

On approaches like these and others, stakeholders can and will "agree to disagree." However, merely discussing issues such as net metering can result in various constituencies moving beyond conflict to consensus to create forward momentum toward realizing the Smart Grid.

HOW NET METERING WORKS IN PENNSYLVANIA

Properly designed regulations & policies like net metering can further the development of the Smart Grid.

In Pennsylvania, investor-owned utilities must offer net metering to residential customers that generate electricity with systems up to 50 kilowatts (kW) in capacity; nonresidential customers with systems up to three megawatts (MW) in capacity; and customers with systems greater than 3 MW but no more than 5 MW who make their systems available to the grid during emergencies. It is available when any portion of the electricity generated is used to offset on-site consumption.

Systems eligible for net metering include those that generate electricity using photovoltaics (PV), solar-thermal energy, wind energy, hydropower, geothermal energy, biomass energy, fuel cells, combined heat and power (CHP), municipal solid waste, waste coal, coal-mine methane, other forms of distributed generation (DG) and certain demand-side management technologies.

Net metering is achieved using a single, bi-directional meter – i.e., two-way communication – that can measure and record the flow of electricity in both directions at the same rate. Net excess generation is carried forward and credited to the customer's next bill at the full retail rate, which includes the generation, transmission and distribution components.

NIST is matching its expertise with DOE's domain expertise to formulate a Smart Grid Roadmap, set to be released by the end of 2009.

STANDARDS & SECURITY: GETTING TO CERTAINTY.

Present and future architects of the Smart Grid look for regulatory certainty before they can confidently enter the marketplace with their respective tools, technologies and deployment plans. Meanwhile, many regulators are seeking evidence of mature interoperability and security standards before they can convey such certainty. Historically, in industries from telecommunications to computers, standards follow markets rather than lead them. That said, standards in both areas are evolving with all deliberate speed.

ABOUT NIST

Founded in 1901, NIST is a non-regulatory federal agency whose mission is to promote U.S. innovation and industrial competitiveness by advancing measurement science, standards, and technology in ways that enhance economic security and improve our quality of life. NIST has created standards for everything from automated teller machines and atomic clocks to mammograms and semiconductors. The agency has been designated within EISA 2007 (Title XIII) to develop the standards framework for Smart Grid technologies.

A status report:

INTEROPERABILITY: NIST AND THE ROADMAP

Many within the grid community argue that waiting for standards is the only way to ensure cost-effective implementation. Others hold that the only standard required is the size of the plug for Smart Grid appliances. Still others maintain that waiting for standards might have retarded the growth of personal computing to the extent that we'd still be playing Pong.

Clearly, there are technologies that can and are being implemented within utilities in anticipation of Smart Grid adoption, among them a wide array of smart sensors. And as long as open technology-neutral standards are observed, private industry is free to develop standards on its own. However, the National Institute of Standards and

Technology (NIST) will draw the Interoperability Roadmap. Ultimately, interoperability standards are needed to ensure that power electronics, communication data and information technology will work together seamlessly, while cyber security standards protects the multi-system network against natural or human-caused disruptions.

NIST is matching its expertise with DOE's domain expertise to formulate a Smart Grid Roadmap, set to be released by the end of 2009. At the same time, the GridWise Architecture Council has begun to develop an interoperability maturity model to determine the appropriate process for developing software.

These efforts provide a starting point to bring the stakeholders together to work toward common goals and visions of what the Smart Grid needs to become.

> A smarter
> grid delivers
> end use conservation and
> efficiency thanks to its ability
> to establish more focused
> and persistent consumer
> participation.

SMART GRID & THE ENVIRONMENT: ENABLING A CLEANER ENERGY FUTURE.

In 2008, emissions of carbon dioxide from fuel burning in the United States were down 2.8%, the biggest annual drop since the 1980s.[10] This is widely attributable to the length and depth of the worldwide recession and just as widely expected to be an anomaly. Most agree, as the national and global economies improve, carbon emissions will resume their upward trend.

ANOTHER POINT IN THE "PLUS COLUMN"

Due to its greater facility for integrating Volt-VAR control, formerly known as conservation voltage reduction, the Smart Grid can minimize losses and resistive loads by continually optimizing distribution system voltage. This can result in systemwide energy savings of approximately 1-2%.

The smarter grid delivers end-use conservation and efficiency thanks to its ability to establish more focused and pervasive consumer participation. From a behavioral perspective, there is measurable energy savings when consumers participate, approximately 6% in the residential sector. Awareness on the part of consumers to manage peak load by virtue of a feedback mechanism incites greater attention to consumption patterns and results in savings.

Proving that timing is everything, a smarter grid can capture carbon savings from peak load shifting — even if energy is not being saved. When peak load is reduced by means of demand response, many peaking plants — and the carbon they emit — are kept on the sidelines.

THE SMART GRID & PLUG-IN ELECTRIC VEHICLES

The Smart Grid's single biggest potential in delivering carbon savings is in providing cost-effective and increasingly clean energy for plug-in electric vehicles (PEVs), including plug-in hybrid electric vehicles (PHEVs). Although the vehicles will be producing the savings rather than the Smart Grid, only Smart Grid technologies will allow us to tap their fundamental potential. The idle production capacity of today's grid could supply 73% of the energy needs of today's cars, SUVs, pickup trucks, and vans with existing power plants.[11] Additional benefits include the potential to displace 52% of net oil imports (or 6.7 million barrels per day) and to reduce CO_2 emissions by 27%.[12]

At scale, PHEV deployment will cut GHG emissions including CO_2. In the process, it will work toward improving the general health of the United States as well as lessening our dependence on foreign oil.

Furthermore, by enabling the sale of more electricity over the same infrastructure, the Smart Grid has the potential to lower electric rates. These benefits accrue, however, only if these vehicles are charged strictly off-peak. Charging PEVs on-peak would only further stress the grid.

In terms of carbon emissions, the nation's vehicles produce roughly the same carbon emissions as the nation's coal-based power plants. By moving their emissions from millions of tailpipes to far fewer smokestacks, the Smart Grid could dramatically reduces the size and complexity of the industry's ongoing "clean-up detail." That is, rather than wondering how to handle hundreds of millions of four-wheeled emitters, Smart-Grid functionality enables us to shift focus to challenges ranging from carbon management to the use of more renewable sources of electricity.

At scale, PHEV deployment will cut GHG emissions including CO_2. In the process, it will work toward improving the general health of the United States as well as lessening our dependence on foreign oil.

ENABLING CARBON SAVINGS

The full exploitation of renewable energy sources such as wind and PV solar is critical to managing our collective carbon footprint. However, when viewed against the limitations of the current grid, both technologies face barriers to full-scale deployment. A smarter grid enables grid operators to see further into the system and allows them the flexibility to better manage the intermittency of renewables. This in turn surmounts a significant barrier, enabling wind and solar to be deployed rapidly — and in larger percentages.

OPTIMIZING WIND

Although possessing myriad attributes, renewables also increase the complexity of operating the grid. A smarter grid enables operators to manage against this complexity.

The Smart Grid can lower the net cost for wind power by regulating fluctuations with demand response. Combining demand response, energy storage and distributed and centralized generation assets can manage

POTENTIAL IMPACTS of HIGH PENETRATION of PLUG-IN HYBRID ELECTRIC VEHICLES on THE US POWER GRID

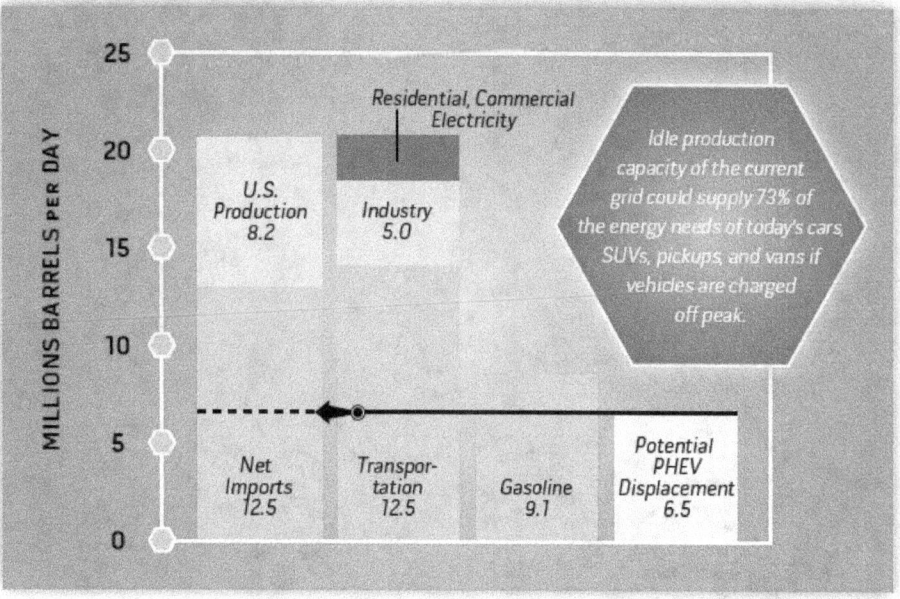

15

CAP & TRADE & SMART GRID

● Congress is working on proposed legislation that would limit greenhouse gas emissions and turn them into a commodity that can be bought and sold (i.e., cap and trade). Accurate accounting of actual carbon footprints made possible by a smarter grid offers solid verification, thereby capturing the value and enhancing the tradability of carbon offsets.

these fluctuations (i.e., when the wind doesn't blow) to lower the cost of integrating wind into the system. Overall, the Smart Grid can optimize the penetration of renewables into our nation's electrical system.

A smarter grid can optimize wind resources in conjunction with demand response controls, dealing with the intermittency of such resources by actively managing "holes in the wind."

OPTIMIZING SOLAR

A PV array on every roof would be a welcome sight. However, although existing distribution grids are capable of safely supporting initial penetrations of PV solar, placing excess power back onto the grid may pose problems. Smart Grid control systems will help the grid rise to this challenge.

ENABLING STORAGE

The Smart Grid enables utilities to put more batteries and other forms of energy storage in more places. Stationed at thousands of points throughout the Smart Grid, they will provide additional electricity resources throughout the system.

SUMMARY OF ENERGY-SAVING AND CARBON-REDUCTION MECHANISMS ENABLED BY THE SMART GRID

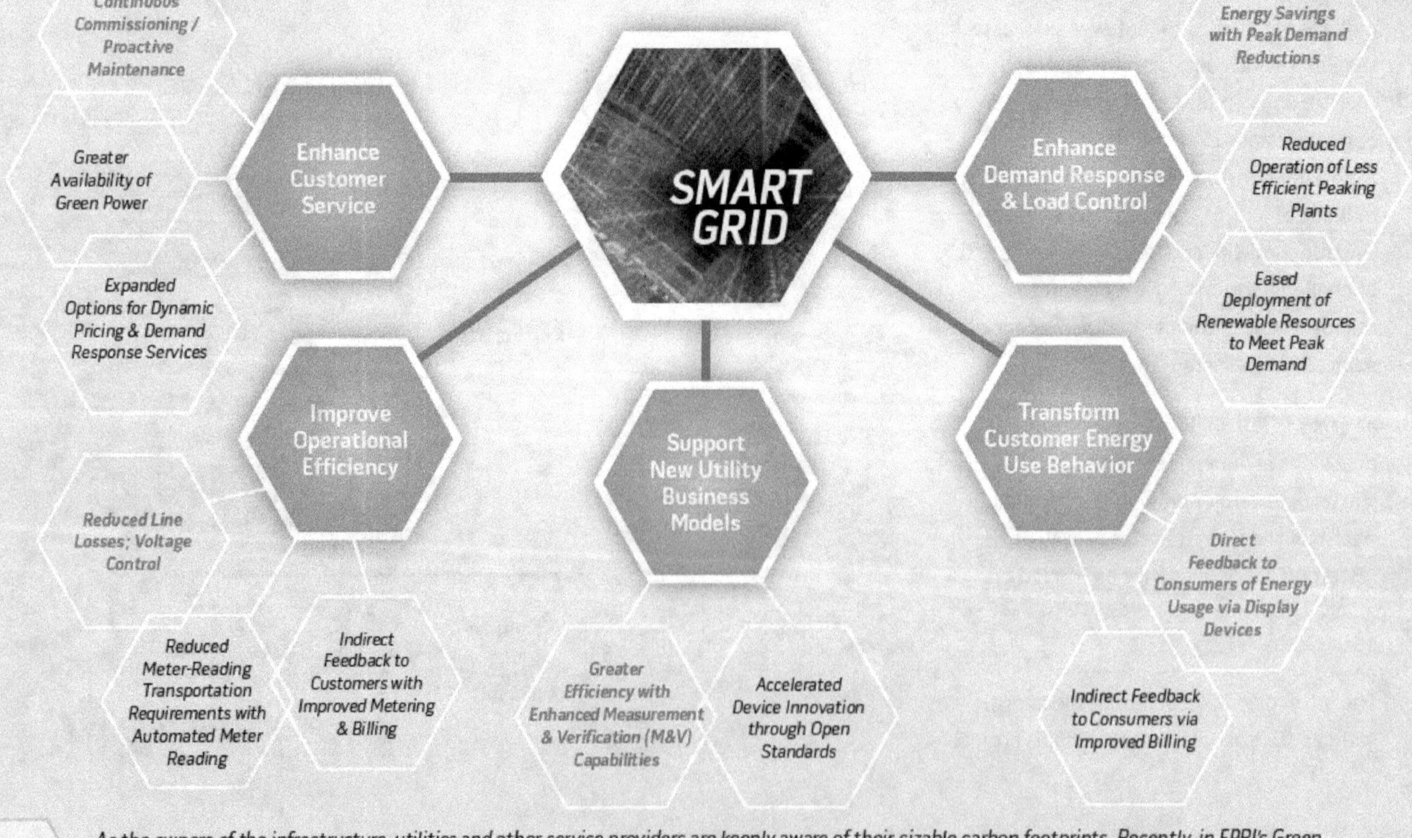

- Continuous Commissioning / Proactive Maintenance
- Greater Availability of Green Power
- Expanded Options for Dynamic Pricing & Demand Response Services
- Reduced Line Losses; Voltage Control
- Reduced Meter-Reading Transportation Requirements with Automated Meter Reading
- Indirect Feedback to Customers with Improved Metering & Billing

SMART GRID

- Enhance Customer Service
- Improve Operational Efficiency
- Support New Utility Business Models
- Enhance Demand Response & Load Control
- Transform Customer Energy Use Behavior

- Energy Savings with Peak Demand Reductions
- Reduced Operation of Less Efficient Peaking Plants
- Eased Deployment of Renewable Resources to Meet Peak Demand
- Direct Feedback to Consumers of Energy Usage via Display Devices
- Indirect Feedback to Consumers via Improved Billing
- Greater Efficiency with Enhanced Measurement & Verification (M&V) Capabilities
- Accelerated Device Innovation through Open Standards

As the owners of the infrastructure, utilities and other service providers are keenly aware of their sizable carbon footprints. Recently, in EPRI's Green Grid Whitepaper, the Institute identified ways in which utilities can reduce carbon through the use of Smart Grid approaches and technologies.

STATES TAKING ACTION:

32 states and the District of Columbia have developed and adopted renewable portfolio standards, which require a pre-determined amount of a state's energy portfolio (up to 20%) to come exclusively from renewable sources by as early as 2013.

STATE	AMOUNT	YEAR	RPS ADMINISTRATOR
Arizona	15%	2025	Arizona Corporation Commission
California	33%	2030	California Energy Commission
Colorado	20%	2020	Colorado Public Utilities Commission
Connecticut	23%	2020	Department of Public Utility Control
District of Columbia	20%	2020	DC Public Service Commission
Delaware	20%	2019	Delaware Energy Office
Hawaii	20%	2020	Hawaii Strategic Industries Division
Iowa	105 MW	–	Iowa Utilities Board
Illinois	25%	2025	Illinois Department of Commerce
Massachusetts	15%	2020	Massachusetts Division of Energy Resources
Maryland	20%	2022	Maryland Public Service Commission
Maine	40%	2017	Maine Public Utilities Commission
Michigan	10%	2015	Michigan Public Service Commission
Minnesota	25%	2025	Minnesota Department of Commerce
Missouri	15%	2021	Missouri Public Service Commission
Montana	15%	2015	Montana Public Service Commission
New Hampshire	23.8%	2025	New Hampshire Office of Energy and Planning
New Jersey	22.5%	2021	New Jersey Board of Public Utilities
New Mexico	20%	2020	New Mexico Public Regulation Commission
Nevada	20%	2015	Public Utilities Commission of Nevada
New York	24%	2013	New York Public Service Commission
North Carolina	12.5%	2021	North Carolina Utilities Commission
North Dakota*	10%	2015	North Dakota Public Service Commission
Oregon	25%	2025	Oregon Energy Office
Pennsylvania	8%	2020	Pennsylvania Public Utility Commission
Rhode Island	16%	2019	Rhode Island Public Utilities Commission
South Dakota*	10%	2015	South Dakota Public Utility Commission
Texas	5,880 MW	2015	Public Utility Commission of Texas
Utah*	20%	2025	Utah Department of Environmental Quality
Vermont*	10%	2013	Vermont Department of Public Service
Virginia*	12%	2022	Virginia Department of Mines, Minerals, and Energy
Washington	15%	2020	Washington Secretary of State
Wisconsin	10%	2015	Public Service Commission of Wisconsin

*Five states, North Dakota, South Dakota, Utah, Virginia, & Vermont, have set voluntary goals for adopting renewable energy instead of portfolio standards with binding targets.

The smart grid will empower average energy consumers to a degree unimaginable just a few years ago.

THE SMART GRID PROGRESS REPORT: **WHO'S DOING WHAT OUT THERE?**

Attempting to gauge the rate of acceptance for a smarter grid reveals a fluid landscape of changing attitudes, successful Smart Grid programs, and appliances that think.

PEOPLE

What will the Smart Grid do for consumers? And how much do consumers care?

In addition to making grid operations as a whole more reliable – an extremely worthy goal in itself – the smart grid will empower average energy consumers to a degree unimaginable just a few years ago. Given new awareness, understanding and tools, they'll be able to make choices that save money, enhance personal convenience, improve the environment – or all three.

Until recently, the overwhelming majority of consumers considered energy a passive purchase. According to conventional wisdom, no one really wanted to think about it. And, frankly, why would they want to? Historically, the system never differentiated the true cost of electricity to the consumer, so they've been programmed not to care. Recent research, however, indicates that this perception has changed significantly. Research conducted in

2007 by Energy Insights indicates that consumers are interested in opportunities afforded them by the Smart Grid.

While some consumers will opt for continued passivity, many more want to be involved in managing how and when they consume energy. Living in a world of seemingly endless customer choice – courtesy of the Internet, telecom and YouTube – consumers have grown impatient with systems characterized by one-way communication and consumption. Research by Energy Insights also reveals that 70% of respondents expressed "high interest" in a unit that keeps them apprised of their energy use as well as dynamic pricing.

Another key trigger for the growth of this consumer class has been growing environmental awareness. A key frustration is that members of this class don't have the tools to make these choices. Once Smart Grid technologies get this information into their hands, customers will enjoy greater levels of

satisfaction and service as measured by outage minutes and have the sense that they can control their bills. More broadly, they'll be able to do their part to reduce peak, which gives rise to both environmental and economic benefits.

PLACES

Austin, Texas

Austin Energy, a utility thoroughly focused on the bottom line due to its municipal ownership, thought it was embarking on a modernization project. Instead, it went far beyond that objective, enabling consumer choice through a wide array of programs including demand response/load management, distributed generation and renewable energy programs. Programs such as these enabled the utility to fund investment in new technologies *at no extra cost to consumers*. Recent deployment included 130,000 smart meters and 70,000 smart thermostats. When transmission and distribution sensors are added, 100% of Austin Energy's consumer base will be served by Smart Grid technologies.

Olympic Peninsula, Washington

One of the first multi-dimensional DOE Smart Grid demonstration projects asked electricity customers to specify a set of simple energy preferences — and then forget about them. In the background, the utility managed energy through smart appliances and thermostats on the customer's behalf, saving customers approximately 10% on average.[13] A true measure of customer acceptance — many didn't want the project to end.

(SMART) THINGS

As for the state of smart appliances, major home-appliance manufacturers are sufficiently convinced of the commercial viability of the Smart Grid.

Whirlpool, the world's largest manufacturer and marketer of major home appliances, has announced that it plans to make all of its electronically controlled appliances Smart Grid compatible by 2015. The company will make all the electronically controlled appliances it produces — everywhere in the world — capable of receiving and responding to signals from the Smart Grid. The company mentioned that its ability to successfully deliver on this commitment in this time frame was dependent on two important public-private partnerships: First, the development by the end of 2010 of an open, global standard for transmitting signals to and receiving signals from a home appliance; and second, appropriate policies that reward consumers, manufacturers and utilities for adding and using these new peak demand reduction capabilities.

GE's smart appliances — or demand-response appliances — include a refrigerator, range, microwave, dishwasher and washer and dryer. Currently running as a pilot program, these appliances receive a signal from the utility company's smart meter, which alerts the appliances — and the participants — when peak electrical usage and rates are in effect. In the pilot program, the signal word "eco" comes up on the display screen. The appliances are programmed to avoid energy usage during that time or operate on a lower wattage; however, participants could choose to override the program.

Moving forward can't be done without adopting a systems view. Utilities in search of a starting place need look no further than the Smart Grid Maturity Model.

THE SMART GRID MATURITY MODEL: *BECAUSE ONE SIZE DOESN'T FIT ALL.*

No two electricity service providers are alike. Nor are their business plans or investment strategies. As utilities across the country consider investing in a Smart Grid, they're also searching for a reasonable degree of solid footing. Utility executives want to know that making the grid smarter is good business with clear benefits.

In effect, how does a Smart Grid-curious utility "do" the Smart Grid?

Moving forward toward the Smart Grid can't be done without adopting a systems view. Utilities in search of a starting place need look no further than the Smart Grid Maturity Model (SGMM). The Maturity Model creates a roadmap of activities, investments and best practices with the Smart Grid as its destination. Utilities using the model will be able to establish an appropriate development path, communicate strategy and vision, and assess current opportunities. The Maturity Model can also serve as a strategic framework for vendors, regulators and consumers who have or desire a role in Smart Grid transformation.

Maturity models – which enable executives to review the progress a business is making in transforming or altering the way it operates – have an admirable track record of moving

entire industries forward. Consider, for example, how they have transformed the software development industry.

During 2007-2009, IBM and seven utilities from four continents developed the Maturity Model and recently donated it to the Carnegie Mellon Software Engineering Institute (SEI). The SEI has developed worldwide de facto standards, such as the Capability Maturity Model Integration (CMMI) for process improvement, and led international efforts to improve network security through its globally recognized Computer Emergency Response Team (CERT) program.

The U.S. Department of Energy is working with the SEI, enabling the Institute to serve as the independent steward of the global SGMM with primary responsibility for its ongoing governance, growth and evolution based upon stakeholder needs, user feedback and market requirements.

LEVEL	ONE: Exploring and Initiating	TWO: Functional Investing	THREE: Integrating Cross Functional	FOUR: Optimizing Enterprise Wide	FIVE: Innovating Next Wave of Improvements
DESCRIPTION	Contemplating Smart Grid transformation. May have vision but no strategy yet. Exploring options. Evaluating business cases, technologies. Might have elements already deployed.	Making decisions, at least at a functional level. Business cases in place, investment being made. One or more functional deployments under way with value being realized. Strategy in place.	Smart Grid spreads. Operational linkages established between two or more functional areas. Management ensures decisions span functional interests, resulting in cross-functional benefits.	Smart Grid functionality and benefits realized. Management and operational systems rely on and take full advantage of observability and integrated control across and between enterprise functions.	New business, operational, environmental and societal opportunities present themselves, and the capability exists to take advantage of them.
RESULT	Vision	Strategy	Systemization	Transformation	Perpetual Innovation

PARTICIPATION TO DATE

To support widespread adoption and use, the SEI will ensure availability of the model and supporting materials and services for the user community, including a suite of offerings on how to use the tool and "train the trainer" sessions.

It is important to note that the Smart Grid Maturity Model is not a means of comparing one utility with another; rather, the intent is strictly one of self-assessment. The first step for utilities is taking the Smart Grid Maturity Model survey by contacting customer-relations@sei.cmu.edu. The survey offers insights into a utility's current position relative to adoption and development of the business plan necessary to set milestones toward achieving the benefits of the Smart Grid — for both residential and business customers.

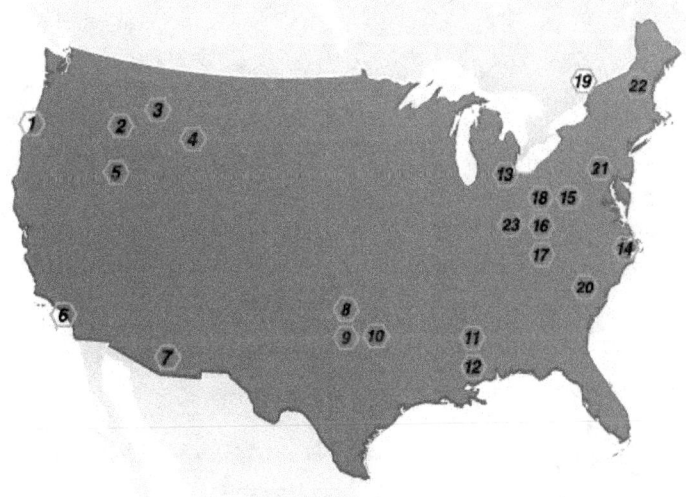

1. PORTLAND GEN.	6. SEMPRA	12. EAST MISS. EPA	18. AEP
2. BC HYDRO	7. SALT RIVER PROJECT	13. COMED	19. HYDRO OTTAWA
3. EPCOR	8. COSERVE	14. DOMINION VIR.	20. SCANA CORP.
4. MANITOBA HYDRO	9. AUSTIN ENERGY	15. ALLEGHENY POWER	21. EXELON
5. BONNEVILLE POWER	10. CENTERPOINT	16. PEPCO	22. VELCO
	11. ENTERGY	17. DUKE	23. FIRST ENERGY

> *Simply put, the purpose of the Collaborative is to get a fix on the state of Smart Grid issues, technologies and best practices.*

FERC, NARUC & THE SMART GRID CLEARINGHOUSE: USING THE POWER OF COLLABORATION TO DRAW CLARITY FROM COMPLEXITY.

SMART GRID "FOR THE REST OF US"

Analogous to the Clearinghouse, the Department of Energy will also launch www.smartgrid.gov. Created for a far broader audience – a "typical" American consumer of electricity interested in the country's energy plan but possibly puzzled by its complexity – this site will keep the public informed about DOE's activities in support of the Smart Grid in an easy-to-understand manner. The site will also function as a single point of entry for the general and trade news media, providing a value-added reference point for this key outreach constituency.

DOE-sponsored Smart Grid projects of various sizes and scope are increasingly coming before regulatory commissions in jurisdictions across the country.

In terms of generating enduring benefits to the grid and society, the Smart Grid represents seven defining and beneficial functions:

- *Accommodating all generation and storage options*

- *Enabling informed participation by customers*

- *Enabling new products, services and markets*

- *Providing the power quality for the range of needs in the 21st century*

- *Optimizing asset utilization and operating efficiently*

- *Addressing disturbances through automated prevention, containment and restoration*

- *Operating resiliently against physical and cyber events and natural disasters*

Clearly, these functions are desirable by any standard. Yet reconciling their value with the day-to-day business before the nation's regulators is complex at best. Regulators are hard at work balancing competing priorities; keeping utility service reliable and affordable; "greening" the electricity supply; modernizing transmission; and combating climate change. Where precisely does the Smart Grid "fit" in their busy schedules and what does it mean to the ratepayers they serve?

FERC/NARUC SMART GRID COLLABORATIVE

To further their understanding with regard to the range of issues associated with the Smart Grid, federal and state regulatory officials have joined together under DOE sponsorship to form the FERC/NARUC Smart Grid Collaborative, using collaboration to draw clarity from complexity.

The Collaborative brings information to regulators so they can get a better sense of the state of Smart Grid issues, technologies and best practices. At joint meetings, regulators interact with a wide array of subject-matter experts on issues that range from AMI to interoperability standards to appropriate timing for Smart Grid deployments. Additionally, they are apprised of Smart Grid projects already at work. Most recently, at the request of the two organizations, DOE has established the Smart Grid Clearinghouse, a comprehensive website built to house "all things Smart Grid," detail and analyze best practices and enable regulators to make more informed ratemaking decisions.

THE SMART GRID CLEARINGHOUSE

The Collaborative sees the Smart Grid Clearinghouse as an additional tool for Smart Grid stakeholders to utilize in advancing Smart Grid concept and implementation as well as a venue for many federal and state agencies and public and private sector organizations to assess Smart Grid development and practices. Public and private entities and their representing associations — collectively referred to as the Smart Grid community — can also benefit from Clearinghouse access. These entities could include, but are not limited to:

• Federal governmental agencies or affiliations (e.g., the U.S. Department of Energy and its Electricity Advisory Committee; the Federal Energy Regulatory Commission, the National Institute of Standards and Technology, and the multi-agency Federal Smart Grid Task Force)

• National Association of Regulatory Utility Commissioners; State regulatory bodies (e.g., public utility or energy commissions)

• Industry or trade associations (e.g., electric utilities, product and service suppliers, Electric Power Research Institute, Edison Electric Institute, National Rural Electric Cooperative Association, American Public Power Association, GridWise Alliance, National Electrical Manufacturers Association)

• End users and many other Smart Grid stakeholders

The Smart Grid Clearinghouse will serve as a repository for public Smart Grid information and direct its users to other pertinent sources or databases for additional public Smart Grid information. The Clearinghouse will become the preeminent resource for stakeholders interested in researching high-level Smart Grid developments and keeping abreast of updates.

In general, the Clearinghouse will be established and maintained in a timely manner that will make data from Smart Grid demonstration projects and other sources available to the public.

To ensure transparency and maximize "lessons learned," recipients of DOE Smart Grid Investment Grants will be required to report setbacks as well as successes on the site. Accentuating such lessons will speed knowledge transfer, facilitate best practices and hasten the progress of all Smart Grid initiatives.

As one of the owners of the infrastructure, utilities must take the leadership role in Smart Grid vision creation and catalyzing stakeholder efforts.

NEXT STEPS: SUMMONING THE ENERGY.

Right now, your utility may be considering committing to a new technology. Moving toward Smart Grid adoption requires that this effort become a full-court press as you consider all Smart Grid technology options rather than a well-publicized few. As the owners of the infrastructure, utilities must take the leadership role in Smart Grid vision creation and catalyzing stakeholder efforts.

In terms of getting started, there's no time like the present.

In 2008, 48% of our nation's trade deficit was a direct result of the petroleum we imported.[15] With global competition for that resource increasing by the hour, our national energy bill will inevitably skyrocket. Utilities that choose to stay the course with our current grid in a world of increasingly diminishing resources — making only incremental improvements when conditions are favorable and funds available — can expect to run out of both choices and time. Without the adoption of the Smart Grid, its principles and its technologies, you will have a limited amount of choices to meet future challenges.

What was once considered "safe" organizational behavior on the part of utilities can now be looked upon as risky. Looking ahead, your utility's capacity to take calculated risks by means of Smart Grid investment is in fact a safer plan.

HOW SHOULD UTILITIES PROCEED? START WITH CONSENSUS BUILDING

Stakeholders — governors, mayors and consumer advocates among them — are beginning to understand and appreciate the Smart Grid in concept and the need to modernize the grid. The problem is often one of economics. Some commissions have no staff and others no budget to attend, for example, FERC/NARUC Collaborative meetings. To the extent that they can, utilities can benefit from spending more time with their commissions and consumer advocates to develop their particular Smart Grid vision together.

Take the lead in local and regional Smart Grid awareness.

Under EISA, DOE is charged with building Smart Grid awareness at the national level. In addition to relying on DOE materials, you have the responsibility for clarifying the application of Smart Grid concepts to your locality.

According to EPRI, by modernizing the grid and implementing Smart Grid technology, the United States can save $638 – $802 billion over 20 years, producing an overall benefit to cost ratio of 4:1 to 5:1. Put another way, this means that every dollar spent on the Smart Grid will produce savings of four to five dollars.

Creating a local awareness program will better and more immediately inform your customers, policymakers and students — the keepers of tomorrow's Smart Grid.

PRACTICE TRANSPARENCY

The idea that Smart Grid investments will not meet current cost-effectiveness tests, while generally the case, will require more explanation rather than less. Take time to make the business case — including the societal business case — as completely as possible.

HEED THE CALL

The U.S. Congress has identified the Smart Grid as a national priority with inherent benefits to our economy and the safety and health of our citizens. Your participation in its adoption will move our nation toward increased security, a cleaner environment and a stronger economy.

The Smart Grid offers utilities choices — while they still have choices to make.

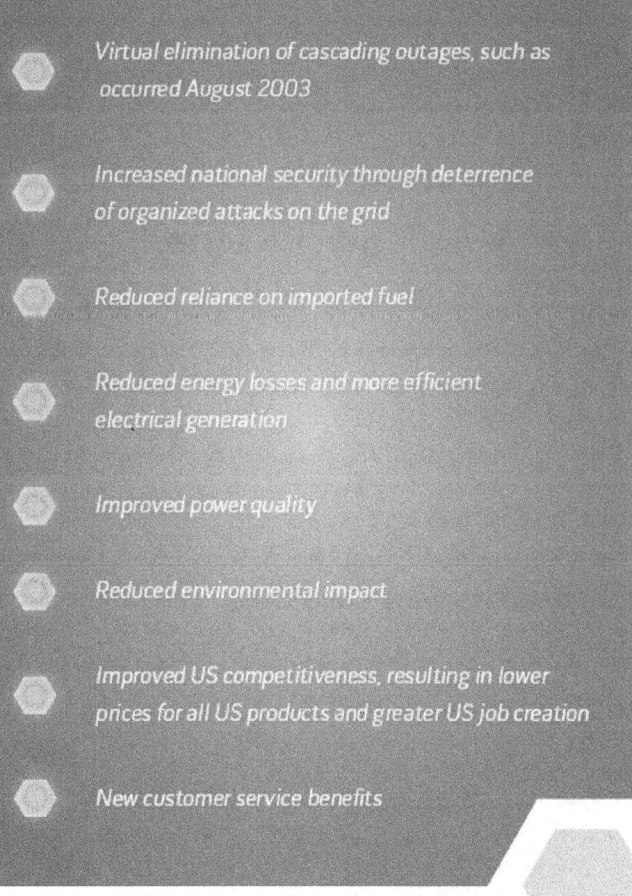

at-a-glance

Benefits
of the Smart Grid:

- Virtual elimination of cascading outages, such as occurred August 2003

- Increased national security through deterrence of organized attacks on the grid

- Reduced reliance on imported fuel

- Reduced energy losses and more efficient electrical generation

- Improved power quality

- Reduced environmental impact

- Improved US competitiveness, resulting in lower prices for all US products and greater US job creation

- New customer service benefits

ADVANCED METERING INFRASTRUCTURE (AMI): AMI is a term denoting electricity meters that measure and record usage data at a minimum, in hourly intervals, and provide usage data to both consumers and energy companies at least once daily.

CARBON DIOXIDE (CO_2): A colorless, odorless, non-poisonous gas that is a normal part of Earth's atmosphere. Carbon dioxide is a product of fossil-fuel combustion as well as other processes. It is considered a greenhouse gas as it traps heat (infrared energy) radiated by the Earth into the atmosphere and thereby contributes to the potential for global warming. The global warming potential (GWP) of other greenhouse gases is measured in relation to that of carbon dioxide, which by international scientific convention is assigned a value of one (1).

DEMAND RESPONSE: This Demand-Side Management category represents the amount of consumer load reduction at the time of system peak due to utility programs that reduce consumer load during many hours of the year. Examples include utility rebate and shared savings activities for the installation of energy efficient appliances, lighting and electrical machinery, and weatherization materials.

DISTRIBUTED GENERATOR: A generator that is located close to the particular load that it is intended to serve. General, but non-exclusive, characteristics of these generators include: an operating strategy that supports the served load; and interconnection to a distribution or sub-transmission system.

DISTRIBUTION: The delivery of energy to retail customers.

ELECTRIC POWER: The rate at which electric energy is transferred. Electric power is measured by capacity.

ELECTRIC UTILITY: Any entity that generates, transmits, or distributes electricity and recovers the cost of its generation, transmission or distribution assets and operations, either directly or indirectly. Examples of these entities include: investor-owned entities, public power districts, public utility districts, municipalities, rural electric cooperatives, and State and Federal agencies.

ENERGY EFFICIENCY, ELECTRICITY: Refers to programs that are aimed at reducing the energy used by specific end-use devices and systems, typically without affecting the services provided. These programs reduce overall electricity consumption (reported in megawatthours), often without explicit consideration for the timing of program-induced savings. Such savings are generally achieved by substituting technologically more advanced equipment to produce the same level of end-use services (e.g. lighting, heating, motor drive) with less electricity. Examples include high-efficiency appliances, efficient lighting programs, high-efficiency heating, ventilating and air conditioning (HVAC) systems or control modifications, efficient building design, advanced electric motor drives, and heat recovery systems.

FEDERAL ENERGY REGULATORY COMMISSION (FERC): The Federal agency with jurisdiction over interstate electricity sales, wholesale electric rates, hydroelectric licensing, natural gas pricing, oil pipeline rates, and gas pipeline certification. FERC is an independent regulatory agency within the Department of Energy and is the successor to the Federal Power Commission.

GREENHOUSE GASES (GHGs): Those gases, such as water vapor, carbon dioxide, nitrous oxide, methane, hydrofluorocarbons (HFCs), perfluorocarbons (PFCs) and sulfur hexafluoride, that are transparent to solar (short-wave) radiation but opaque to long-wave (infrared) radiation, thus preventing long-wave radiant energy from leaving Earth's atmosphere. The net effect is a trapping of absorbed radiation and a tendency to warm the planet's surface.

LOAD (ELECTRIC): The amount of electric power delivered or required at any specific point or points on a system. The requirement originates at the energy-consuming equipment of the consumers.

OFF PEAK: Period of relatively low system demand. These periods often occur in daily, weekly, and seasonal patterns; these off-peak periods differ for each individual electric utility.

ON PEAK: Periods of relatively high system demand. These periods often occur in daily, weekly, and seasonal patterns; these on-peak periods differ for each individual electric utility.

OUTAGE: The period during which a generating unit, transmission line, or other facility is out of service.

PEAK DEMAND OR PEAK LOAD: The maximum load during a specified period of time.

PEAKER PLANT OR PEAK LOAD PLANT: A plant usually housing old, low-efficiency steam units, gas turbines, diesels, or pumped-storage hydroelectric equipment normally used during the peak-load periods.

RATEMAKING AUTHORITY: A utility commission's legal authority to fix, modify, approve, or disapprove rates as determined by the powers given the commission by a State or Federal legislature.

RATE OF RETURN: The ratio of net operating income earned by a utility is calculated as a percentage of its rate base.

RATES: The authorized charges per unit or level of consumption for a specified time period for any of the classes of utility services provided to a customer.

RENEWABLE ENERGY RESOURCES: Energy resources that are naturally replenishing but flow-limited. They are virtually inexhaustible in duration but limited in the amount of energy that is available per unit of time. Renewable energy resources include: biomass, hydro, geothermal, solar, wind, ocean thermal, wave action, and tidal action.

SOLAR ENERGY: The radiant energy of the sun, which can be converted into other forms of energy, such as heat or electricity.

TIME-OF-DAY PRICING: A special electric rate feature under which the price per kilowatthour depends on the time of day.

TIME-OF-DAY RATE: The rate charged by an electric utility for service to various classes of customers. The rate reflects the different costs of providing the service at different times of the day.

TRANSMISSION (ELECTRIC): The movement or transfer of electric energy over an interconnected group of lines and associated equipment between points of supply and points at which it is transformed for delivery to consumers or is delivered to other electric systems. Transmission is considered to end when the energy is transformed for distribution to the consumer.

WIND ENERGY: Kinetic energy present in wind motion that can be converted to mechanical energy for driving pumps, mills, and electric power generators.

DATABASE OF STATE INCENTIVES FOR RENEWABLES & EFFICIENCY (DSIRE): http://www.dsireusa.org

EDISON ELECTRIC INSTITUTE (EEI): http://www.eei.org

ELECTRICITY ADVISORY COMMITTEE (EAC): http://www.oe.energy.gov/eac.htm

ENERGY FUTURE COALITION: http://www.energyfuturecoalition.org

EPRI INTELLIGRID: http://intelligrid.epri.com/

FERC/NARUC COLLABORATIVE: http://www.naruc.org/ferc/default.cfm?c=3

GRID WEEK: http://www.gridweek.com

GRIDWISE ALLIANCE: http://www.gridwise.org

NATIONAL ELECTRICAL MANUFACTURERS ASSOCIATION (NEMA): http://www.nema.org

NATIONAL ENERGY TECHNOLOGY LABORATORY (NETL): http://www.netl.doe.gov/

PACIFIC NORTHWEST NATIONAL LABORATORY (PNNL): http://www.pnl.gov/

PNNL GRIDWISE: http://www.gridwise.pnl.gov/

SMART GRID: http://www.oe.energy.gov/smartgrid.htm

SMART GRID MATURITY MODEL (SGMM): http://www.sei.cmu.edu/smartgrid

SMART GRID TASK FORCE: http://www.oe.energy.gov/smartgrid_taskforce.htm

ENDNOTES

[1] EIA, 2009 Energy Outlook

[2] Smart Grid: Enabling the 21st Century Economy, DOE Modern Grid Strategy, December 2008

[3] ABC News/Washington Post poll, April 30, 2009

[4] Smart Grid: Enabling the 21st Century Economy, DOE Modern Grid Strategy, December 2008

[5] Electricity Advisory Committee, "Smart Grid: Enabler of the New Energy Economy," December 2008

[6] Smart Grid Benefits, DOE Modern Grid Strategy, August 2007

[7] Smart Grid Benefits, DOE Modern Grid Strategy, August 2007

[8] Smart Grid Benefits, DOE Modern Grid Strategy, August 2007

[9] Smart Grid Benefits, DOE Modern Grid Strategy, August 2007

[10] EIA, U.S. Carbon Dioxide Emissions from Energy Sources 2008 Flash Estimate, May 2009

[11] Pacific Northwest National Laboratory, "The Smart Grid and Its Role in a Carbon-constrained World," February 2009

[12] Pacific Northwest National Laboratory, "The Smart Grid and Its Role in a Carbon-constrained World," February 2009

[13] Pacific Northwest National Laboratory, "The Smart Grid and Its Role in a Carbon-constrained World," February 2009

[14] Pacific Northwest National Laboratory, "The Smart Grid and Its Role in a Carbon-constrained World," February 2009

[15] IPAA Access Direct, July 2009